Effects of Skin Contact with Chemicals

Guidance for Occupational Health Professionals and Employers

DEPARTMENT OF HEALTH AND HUMAN SERVICES
Centers for Disease Control and Prevention
National Institute for Occupational Safety and Health

Effects of Skin Contact with Chemicals

Guidance for Occupational Health Professionals and Employers

DEPARTMENT OF HEALTH AND HUMAN SERVICES
Centers for Disease Control and Prevention
National Institute for Occupational Safety and Health

This document is in the public domain and may be freely copied or reprinted.

Disclaimer

Mention of any company or product does not constitute endorsement by the National Institute for Occupational Safety and Health (NIOSH). In addition, citations to Web sites external to NIOSH do not constitute NIOSH endorsement of the sponsoring organizations or their programs or products. Furthermore, NIOSH is not responsible for the content of these Web sites. All Web addresses referenced in this document were accessible as of the publication date.

Ordering Information

To receive documents or other information about occupational safety and health topics, contact NIOSH at

Telephone: **1–800–CDC–INFO** (1–800–232–4636)
TTY: 1–888–232–6348
E-mail: cdcinfo@cdc.gov

or visit the NIOSH Web site at **www.cdc.gov/niosh**.

For a monthly update on news at NIOSH, subscribe to *NIOSH eNews* by visiting **www.cdc.gov/niosh/eNews**.

DHHS (NIOSH) Publication No. 2011–200

August 2011

SAFER • HEALTHIER • PEOPLE™

Contents

INTRODUCTION	1
Scope of Chemical Exposures	2
Skin	3
ADVERSE HEALTH IMPACTS OF CHEMICAL EXPOSURES ON SKIN	4
Adverse Impacts	4
TYPES OF ADVERSE EFFECTS	5
Direct—At Point of Contact	5
Systemic—Away from Site of Entry	5
Major Types of Adverse Effects: Summary	6
EXAMPLES OF DIRECT EFFECTS	7
Defatting/Dry Skin	7
Irritation	7
Corrosion	8
Changes in Pigmentation	8
Chloracne	9
Skin Cancer	9
EXAMPLES OF SYSTEMIC EFFECTS	10
Specific Organs	10
Human Body Systems	10
EXAMPLES OF SENSITIZATION EFFECTS	11
Airway Sensitization	11
EXAMPLES OF COMBINED EFFECTS	12
ASSESSING AND MANAGING RISK OF CHEMICAL EXPOSURES	13
Step by Step	13
Managing Risk Through Prevention and Control	14
FINDING INFORMATION ON CHEMICALS, SKIN EXPOSURES, AND RISK MANAGEMENT	15
Labels	15
NIOSH Hazard-Specific Skin Notations (SK)	15
Hazard-Specific Skin Notations	16
Material Safety Data Sheets (MSDSs)	17
Safety and Health Professionals	17
Learn about available resources	
REMEMBER: S - K - I - N	19

Introduction

Chemical exposure in the workplace is a significant problem in the United States. More than 13 million workers in the United States are potentially exposed to chemicals via the skin. Skin disorders are among the most frequently reported occupational illnesses, resulting in an estimated annual cost in the United States of over $1 billion. While the rates of most other occupational diseases are decreasing, skin disease rates are actually increasing.

Efforts to reduce or prevent skin problems in many work settings are lacking as too frequently workers, employers, and even occupational health professionals accept skin problems as part of the job. The tolerance of occupational skin problems must be lowered and the methods for assessing and reducing chemical exposures must be improved. As occupational health professionals or employers, it is important that you know how to identify and manage the risk of chemical exposures to the skin and prevent injury and illness associated with dermal exposure risks.

This pamphlet will provide occupational health professionals and employers with:

- knowledge of the major adverse health effects resulting from chemical exposures to the skin,

- information on recognizing chemical hazards,

- knowledge of intervention/prevention strategies, and

- sources of information related to skin disorders and prevention.

Scope of Chemical Exposures

Chemical exposures are the main cause of work-related skin disorders. Chemical exposures to the skin are an everyday occurrence for many workers in a wide variety of occupations, including:

- Agriculture
- Manufacturing
- Services
- Transportation/Utilities
- Construction
- Sales

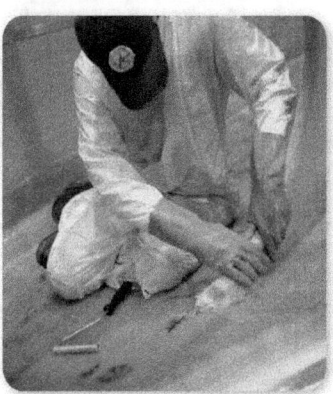 

Skin

Skin is the body's protective cover and the principle site of interaction with the world around it. The skin limits the loss of water and other requisite compounds from the body. At the same time, the skin also limits unwanted substances, including chemicals, from entering the body.

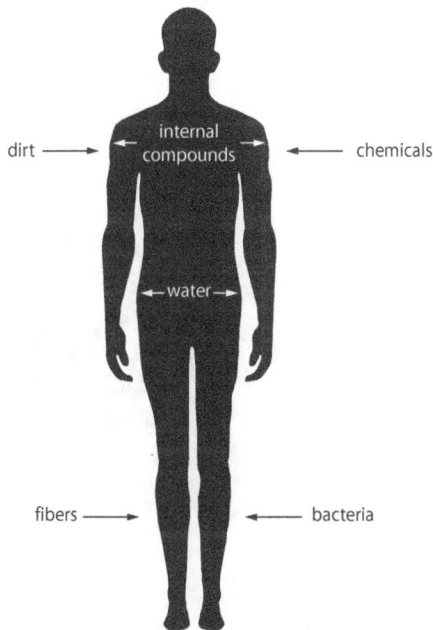

Damage to the skin reduces its ability to protect the body.

Adverse Health Impacts of Chemical Exposures on Skin

Chemical exposures to the skin can result in either temporary or permanent adverse health impacts. These health impacts may occur at the point of contact with the chemical, or the chemical may enter the body through compromised skin (such as a wound) or by permeating the skin. Then the chemical can be distributed by the bloodstream, causing or contributing to a health problem somewhere else in the body.

Adverse Impacts

Temporary

Temporary adverse health impacts may occur from exposure to chemicals. For example, it is not uncommon to experience dry, red, cracked skin from contact with water, soaps, gasoline, and certain types of solvents. These disorders usually heal quickly when the skin is no longer in contact with the chemical, but they can increase the chance of infection when breaks in the skin are present.

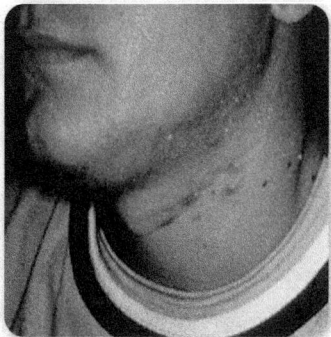

Images illustrating skin injuries and diseases were selected from *Occupational Dermatoses—A Program for Physicians*. Additional information about this slideshow is available at http://www.cdc.gov/niosh/topics/skin/occderm-slides/ocderm.html.

Permanent

Permanent adverse health impacts may result if the skin experiences exposures to a chemical that can cause severe damage. For example, a chemical burn may leave a permanent scar. Exposure to certain chemicals can result in permanent loss of skin color. Permanent damage may also occur to body organs or systems as a result of chemical exposure to the skin.

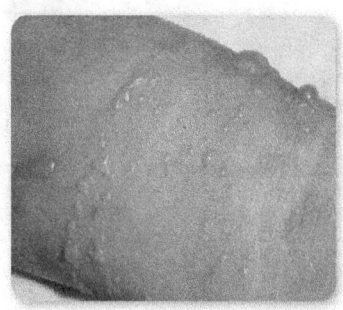

Types of Adverse Effects

Direct—At Point of Contact
A chemical may cause a problem at the point of contact with the body. For example, exposure to a chemical on the wrist rest of a computer keyboard caused the response on this worker's hand.

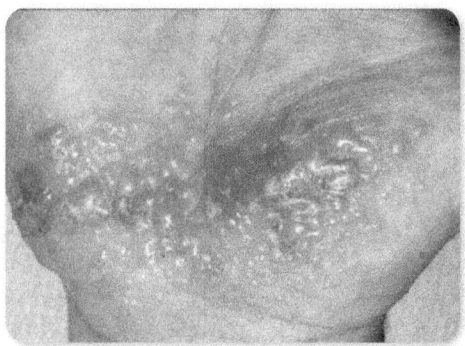

Systemic—Away from Site of Entry
A chemical may enter the body either through intact or compromised (broken) skin and cause or contribute to a health problem somewhere else in the body. Exposure to organophosphate pesticides, which can enter the body through the skin, may cause damage to the nervous system.

Sensitization is another type of health effect resulting from dermal exposure to chemicals. **Combined health effects** from a single chemical exposure may also occur.

Major Types of Adverse Effects: Summary

1. **Direct**—Exposure to chemicals can cause effects at the point of contact. These are called direct effects and include defatting/drying, irritation, corrosion, changes in pigmentation, chloracne, and skin cancer.

2. **Systemic**—Chemicals can enter the body and cause or contribute to health problems somewhere else in the body. These are called systemic effects and may affect a specific organ or an entire body system.

3. **Sensitization**—Chemicals may cause a sensitization effect, in which an individual becomes unusually susceptible to a chemical or group of chemicals. From then on, exposure to even very small amounts of the substance can cause an allergic reaction. The only way to deal with sensitization is to prevent any further exposure or contact with the chemical. Sensitization effects include allergic contact dermatitis and airway sensitization.

4. **Combined**—Chemical exposure to the skin may cause multiple health effects in an exposed individual.

Examples of Direct Effects

Defatting/Dry Skin
Defatting or drying results when a chemical removes the natural oils from the skin. The most frequent causes of defatted or dry skin are exposures to soaps, solvents, and moisture. This effect is temporary if the exposure ceases.

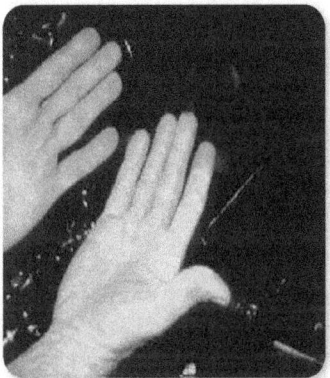

Irritation
Chemical substances that cause reddening, dryness, and cracking of the skin on contact are known as irritants. Irritation is most frequently caused by fiberglass, food products, oils and lubricants, and solvents. If caught in time, no permanent effects occur.

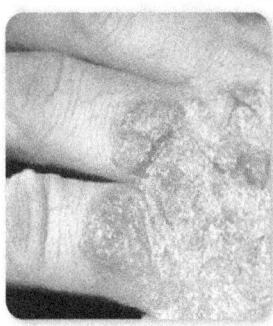
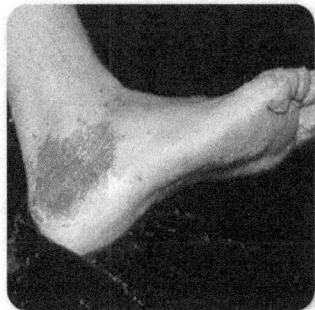

Corrosion

Corrosive substances result in more severe or serious damage to the skin. A chemical burn can result from a brief exposure to a corrosive substance. Corrosive substances include strong alkali (basic) materials and acids. Scarring is a common outcome. Effects resulting from exposure to corrosives are permanent.

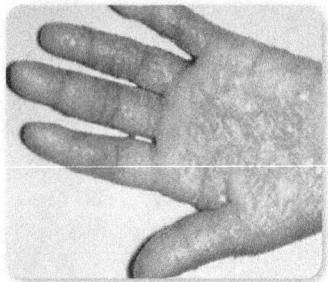

Changes in Pigmentation

A permanent change in skin color may result from exposure to certain chemicals, such as tar and asphalt products and some disinfectants.

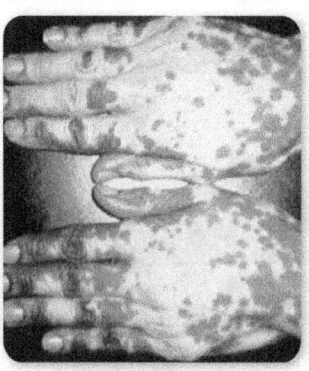

Chloracne

Chloracne is a type of acne caused by certain halogenated aromatic chemicals. It can occur after exposure to polychlorinated hydrocarbons (PCBs) and certain pesticides.

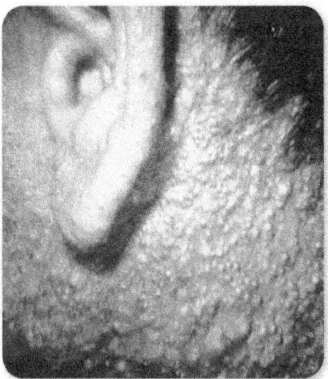

Skin Cancer

Malignant skin tumors may occur after exposure to an occupational carcinogen.

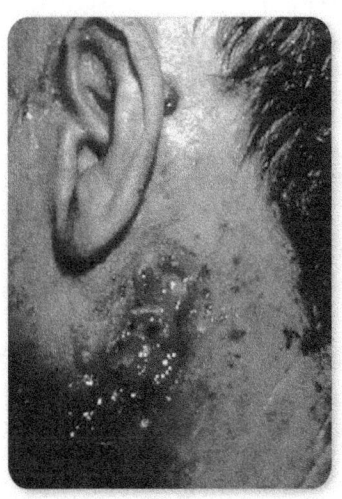

Some chemicals found in the workplace contain cancer-producing substances (carcinogens). When these come in contact with the skin, a malignant tumor may form at the site of contact. Exposure to coal tar resulted in the skin tumor pictured below.

Examples of Systemic Effects

Specific Organs

Chemicals absorbed into the body through the skin can cause damage to a specific organ, such as the liver, kidney, or bladder. Liver and kidney damage may occur from exposure to solvents such as toluene and xylene.

Human Body Systems

Chemicals absorbed through the skin can damage an entire body system, including the immune system, nervous system, or respiratory system. Pesticides and herbicides are chemicals that may impact body systems.

Examples of Sensitization Effects

Allergic contact dermatitis is an allergic response (immunological response) of the skin as a result of exposure to a chemical. Chemical exposures that may result in allergic contact dermatitis include epoxy resins, chromates, rubber chemicals, amine hardeners, and phenol-formaldehyde resins.

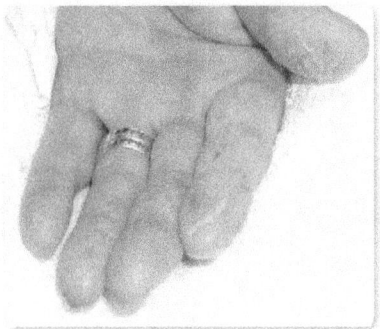

Airway Sensitization

An allergic reaction of the mucous membranes or airways may result when the skin is exposed to certain chemicals. Skin exposure, as well as inhalation of isocyanates (contained in many paints and other building materials, like spray-on insulation and roofing materials), can result in airway sensitization.

Examples of Combined Effects

Chemical exposure to the skin may cause multiple health effects. For example, individuals working with cement may experience combined adverse health effects. Contact with the cement may result in irritation at the point of contact from the alkaline nature of the cement. Workers may also become sensitized to cement due to the chrome salts present in the material.

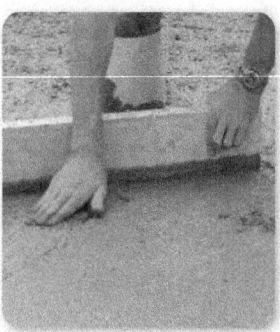 

Work practices that can lead to sensitization and irritation

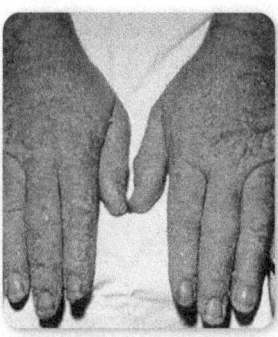 

Sensitization   Irritation

Assessing and Managing Risk of Chemical Exposures

With the knowledge that chemical exposures to skin at work have the potential to cause adverse health effects that are preventable, the next step is assessing the risk of workers developing either temporary or permanent skin-related disorders.

Step by Step

1. To assess the risk for skin-related problems, first identify the chemicals used in the workplace. Do not neglect to consider chemicals that may be generated during the process.

2. Next, identify those chemicals that pose a health risk through skin contact.

3. Assess potential for skin exposure, considering the likelihood that chemicals can be absorbed into the body and the potential for direct skin contact with chemical agents in any form (vapor, liquid, etc.).

4. Implement a skin hazard prevention program, following the hierarchy of controls described in the next section.

5. Document and monitor the presence of skin-related problems through health surveillance activities.

    — Establish a workplace surveillance program.

    — Conduct medical screening.

    — Conduct biological monitoring.

Managing Risk Through Prevention and Control

To reduce the risk of adverse health effects from chemical exposures to the skin, the exposure must be either prevented or controlled.

Prevention

Exposure to harmful chemicals may be prevented by these actions.

Eliminate

Eliminate unnecessary chemicals from a work process. For example, use disposable brushes rather than cleaning them with a solvent.

Substitute

Replace a harmful chemical or product with one that is less harmful. For example, substitute a solvent-based product with a water-based one.

Control

Chemical exposure to the skin can be controlled or reduced by these actions.

Modify the process

Modify a process to eliminate chemical exposure. For example, rather than hand-cleaning metal parts during repair operations, use a mechanical cleaner.

Add ventilation

Reduce airborne exposures by adding local or general ventilation. For example, use ventilation during spray-painting operations to reduce airborne levels of isocyanates.

Maintain healthy skin

Clean skin with mild soap, rinse thoroughly, and use moisturizer. Dry skin is damaged and more affected by chemicals.

Modify work practices

Modify work practices to reduce or eliminate skin contact with chemicals. For example, rather than applying a solvent with a rag, use a brush.

Follow good housekeeping practices

A clean work area helps prevent skin contact with chemicals from work surfaces.

Use personal protective equipment

Use personal protective equipment when exposure to chemicals is unavoidable. This may include chemical-resistant gloves, aprons, coveralls, and boots. For example, use appropriate gloves when mixing epoxy resin to avoid skin contact. Selection of the correct protective equipment is critical. Check a source such as the *Quick Selection Guide to Chemical Protective Clothing*.

Finding Information on Chemicals, Skin Exposures, and Risk Management

Labels

Read labels to identify the chemical contents of materials being used and to be aware of any handling or health warnings.

NIOSH Hazard-Specific Skin Notations (SK)

NIOSH has developed a new system for the assignment of multiple hazard-specific skin notations (SK) to help workers and occupational health professionals understand the health risks of skin exposures to hazardous chemicals. The hazard-specific SK (see table on next page) will appear in future NIOSH publications, including the *NIOSH Pocket Guide to Chemical Hazards*, and will identify the major health effects associated with skin contact.

Chemicals may be assigned more than one hazard-specific SK when they are identified to cause multiple adverse health effects following skin contact. For example, if a chemical is identified as corrosive and also contributes to systemic toxicity, it will be labeled as SK: SYS-DIR (COR).

Hazard-Specific Skin Notations

Hazard-specific SK	Health effects	Definition
SYS	Systemic effects	Chemicals may cause systemic damage to areas of the body that are beyond the site of skin contact.
		Damaged areas may include specific organs (e.g., liver, heart, and kidneys) and biological systems (e.g., nervous system, reproductive system, and immune system).
		Subnotation of SK: SYS
(FATAL)	Lethal or life-threatening effects	Assigned to chemicals identified as highly or extremely toxic that may be potentially lethal or life-threatening following exposure of the skin
DIR	Direct (localized) effects	Chemicals may cause damage at or near the site of skin contact. Health effects may include skin irritation and corrosion, bleaching or darkening of the skin, or skin cancers.
(IRR)	Skin irritation	Subnotation of SK: DIR
(COR)	Skin corrosion	Subnotation of SK: DIR
SEN	Allergic and other immune-mediated reactions	Chemical may cause allergic contact dermatitis (ACD), sensitization of exposed skin, or airway sensitization following skin contact.

Material Safety Data Sheets (MSDSs)

An MSDS is designed to provide workers, emergency personnel, and health care professionals with the proper procedures for handling or working with a particular substance. MSDSs contain information such as physical data, adverse health effects, first aid, reactivity, storage, disposal, protective equipment, and spill/leak procedures.

Safety and Health Professionals

Discuss chemical effects on the skin with knowledgeable safety and health personnel.

Learn about available resources

The Center to Protect Worker's Rights: Electronic Library of Construction Occupational Safety and Health at www.cdc.gov/elcosh

Indexed Dermal Bibliography at (1995–2007) www.cdc.gov/niosh/docs/2009-153/

Current Intelligence Bulletin 61: A Strategy for Assigning New NIOSH Skin Notations at www.cdc.gov/niosh/docs/2009-147/

NIOSH's Skin Exposures and Effects topic page at www.cdc.gov/niosh/topics/skin

NIOSH Pocket Guide to Chemical Hazards and Web site at www.cdc.gov/niosh/npg.html

Quick Selection Guide to Chemical Protective Clothing Fourth Edition at www.cdc.gov/niosh/ncpc1.html

OSHA Dermal Exposure topic page at www.osha.gov/SLTC/dermalexposure/index.html

State and local health departments, such as the Washington Department of Labor and Industries, at www.lni/wa/gov/sharp

British Patient Information at www.patient.co.uk/showdoc/23068731 or www.bad.org.uk/public/leaflets/ContactDermatitis_update_2007.pdf

Remember: S - K - I - N

Social handicap — Skin problems can affect both you and your family's quality of life.

Knowledge — Information is the key to prevention and the cure to skin-related health problems. Help is available.

Impediment — Skin problems cause physical discomfort, limitations on your daily life, loss of personal and work time, and possibly the loss of your job.

Not necessary — A skin-related health problem is not a requirement of your job.

www.ingramcontent.com/pod-product-compliance
Lightning Source LLC
Chambersburg PA
CBHW070736180526
45167CB00004B/1782